Electric Life Force
Fibonatti Publishing

Copyright

Copyright © 2024 by Fibonatti Publishing
All rights reserved.
No portion of this book may be reproduced in any form without written permission from the publisher or author, except as permitted by U.S. copyright law.

Contents

Copyright 2
Introduction 6
    Why Electricity is Fundamental to Life 6
    The Intriguing Connection Between Lightning, Evolution, and Consciousness .. 6
    Overview of the Book's Journey: From the Origins of Life to the Mysteries of the Human Mind 7
Part 1: The Birth of Life 8
Chapter 1: Lightning and the Origin of Life 9
    Early Earth Conditions: A Volatile, Electrically Active Environment 9
    The Miller-Urey Experiment: How Lightning May Have Sparked Amino Acids and the Building Blocks of Life 9
    The Role of Natural Electric Discharges in Catalyzing Complex Organic Molecules 11
Chapter 2: The Earth's Electric Circuit 13
    The Global Electrical System: Earth, Atmosphere, and Ionosphere 13
    How Atmospheric Electricity Influenced Early Biological Systems 14
    Potential Links Between Earth's Natural Electric Fields and Early Microbial Evolution 15
Part 2: Electricity and Evolution 17
Chapter 3: Electricity in Plant Life 18
    How Plants Generate and Use Bioelectric Signals for Communication, Growth, and Defense 18
    Electrical Signaling in Roots and Leaves: the "Nervous System" of Plants 19
    The Role of Plant Bioelectricity in Shaping Ecosystems 20
Chapter 4: Bioelectricity in Animal Evolution 22
    Early Multicellular Organisms and the Rise of Bioelectric Communication 22
    The Evolution of Nervous Systems: Why Electricity Became Central to Animal Function 22
    Specialized Creatures: Electric Fish and the Evolutionary Advantage of Electricity 23
Chapter 5: The Role of Electricity in Human Evolution 24
    How Bioelectricity Influenced Brain Development and Complexity 24
    The Interplay of Diet, Bioelectricity, and Energy Demands in Humans 24

Possible Connections Between Electricity and the Evolution of Human Consciousness 25

Part 3: Electricity and the Human Body 26

Chapter 6: How Electricity Powers Life..27

    Recap of Human Bioelectricity: Cells, Tissues, and Organ Systems 27

    The Role of Ion Channels and Action Potentials in the Nervous System 27

    Electrical Activity in the Heart, Muscles, and Other Tissues 27

Chapter 7: Healing and Regeneration 29

    Natural Electric Fields in Wound Healing and Tissue Repair............................29

    Advances in Bioelectric Medicine: Electroceuticals and Regeneration 29

    Speculative Theories about Limb and Organ Regeneration 30

Chapter 8: The Schumann Resonance and the Human Body 32

    Earth's Natural Resonances and their Potential Connection to Human Biology 32

    Theories Linking the Schumann Resonance to Brain Activity and Circadian Rhythms...32

    Separating Science from Speculation 33

Part 4: Electricity and Consciousness..34

Chapter 9: Brainwaves and the Electric Mind 35

    How the Brain Generates and Uses Electrical Signals to Create Thought and Emotion..35

    The Relationship Between Brainwave Frequencies and States of Consciousness 35

    Advances in Neuroscience: Mapping the Brain's "Electric Landscape" 36

Chapter 10: Electricity and Human Potential 37

    Exploring Electricity's Role in Creativity, Problem-Solving, and Meditative States..37

    The Potential of Electrical Brain Stimulation to Enhance Cognition or Treat Disorders..37

    Speculations About How Bioelectricity Shapes Individuality and Personality..38

Chapter 11: Electricity and the Life Force 40

    Perspectives from Ancient Traditions: Electricity as Life Force.......................40

    Bioelectromagnetism and the Concept of a "Biofield"......................................40

    Speculations about Electricity as a Bridge between Science and Spirituality....40

Chapter 12: Ancient Cultures and the Electric Connection....................................42

Egyptians: Electric Fish and Healing Practices; the Concept of the "Ka" as Life Force 42

Chinese: Qi and Acupuncture; Parallels to Bioelectric Pathways 42

Greeks: Thales and Static Electricity; Early Theories of the Soul as an Energetic Force 42

Indians: Prana and Chakras as Possible Analogs for Bioelectric Energy 43

Native Americans and Mesoamericans: Reverence for Lightning as a Creative Force 43

Electric Rays and Healing: Roman and Islamic use of Electric Fish for Medical Treatment 43

How these Ancient Beliefs and Practices Align with Modern Discoveries 44

Part 5: The Future of Electricity and Life 45

Chapter 13: Bioelectricity in Medicine and Technology 46

Advances in Bioelectric Engineering: Artificial Organs and Regenerative Medicine 46

Using Electricity to Combat Disease 46

Emerging Technologies Inspired by Biological Electricity 47

Chapter 14: Electricity and Artificial Intelligence 49

How Studying Human Bioelectricity Informs AI and Robotics 49

Speculative Futures: Integrating Human Bioelectricity with Machines 49

Conclusion 51

Revisiting the Journey 51

How Understanding Electricity can Transform Medicine, Technology, and Philosophy 51

Final Thoughts 52

Introduction

Why Electricity is Fundamental to Life

Electricity is not just a force confined to technology; it is a phenomenon deeply intertwined with life itself. It powers not only the tools of modern civilization but also the very essence of living organisms. From the smallest bacterium to the complexity of the human body, electricity drives the processes that make life possible.

Electricity is the movement of charged particles—a dynamic, invisible energy that governs everything from the communication between cells to the contraction of muscles and the beating of a heart. In your body, every thought, movement, and heartbeat is powered by electrical signals. These signals, carried by ions like sodium and potassium, create a constant flow of communication within and between cells, ensuring the coordination of life's intricate systems.

Beyond humans, nature showcases electricity in remarkable ways. Electric eels use specialized cells to generate shocks for defense and hunting. At the same time, plants like the Venus flytrap rely on electrical impulses to capture their prey. Even at the dawn of life, electricity may have played a vital role. The raw power of lightning striking Earth's primordial oceans is thought to have triggered the chemical reactions that gave rise to the building blocks of life.

Understanding electricity's role in life is more than just a scientific curiosity; it opens pathways to self-improvement. By recognizing that your body and mind operate as an electrical system, you can make informed choices to enhance your well-being. Practices like meditation, exercise, and quality sleep can optimize the electrical rhythms of your brain and body, boosting focus, reducing stress, and improving resilience.

Electricity is fundamental to life, not just as a physical phenomenon but as a unifying force that connects the origins of existence to the complexities of consciousness. As we embark on this journey, we will uncover how electricity shapes life, evolution, and even the way we understand ourselves.

The Intriguing Connection Between Lightning, Evolution, and Consciousness

Imagine a stormy primordial Earth, its skies alight with bolts of lightning crackling across the heavens. This was no ordinary spectacle; it was the prelude to life as we know it. Lightning, with its immense electrical power, is believed to have acted as a catalyst for the chemical reactions that formed amino acids, the essential building blocks of proteins. These proteins would eventually assemble into the complex molecular machinery required for life.

Scientists have long hypothesized that this dramatic interplay between energy and matter was a turning point in Earth's history. Lightning strikes, combined with the volatile chemical environment of early Earth, likely created the conditions necessary for the first self-replicating molecules to emerge. This was not just a spark in a physical sense but the literal spark of life.

Fast-forward billions of years and electricity continued to play a pivotal role in evolution. As life became more complex, so did its reliance on bioelectricity. Early organisms developed rudimentary electrical systems to communicate, sense their environment, and coordinate their movements. These systems laid the groundwork for the sophisticated nervous systems seen in higher animals, including humans. But how does this connect to consciousness? The human brain, often described as the most complex structure in the universe, operates entirely on electrical impulses. Neurons communicate through action potentials—brief spikes of electrical activity—that enable thought, emotion, and awareness. This intricate web of electrical signals gives rise to the phenomenon we call consciousness. It is the culmination of billions of years of evolution, starting with a single lightning bolt.

Understanding this connection isn't just about looking backward; it's about seeing the potential within ourselves. Just as lightning sparked the beginnings of life, we can learn to harness our own inner electricity—our thoughts, actions, and energy—to shape our futures. From the ancient chaos of a lightning storm to the quiet hum of our neural networks, electricity has been and continues to be the bridge between the physical and the profound.

Overview of the Book's Journey: From the Origins of Life to the Mysteries of the Human Mind

This book takes you on an electrifying journey through life's connection to electricity, revealing how this invisible force shapes everything from the origins of life to the intricacies of human consciousness. We begin by exploring the raw power of nature, delving into how lightning may have sparked life on Earth billions of years ago. From there, we trace the role of electricity in the evolution of organisms, showcasing its importance in plants, animals, and eventually humans.

As we delve deeper, we'll uncover how electricity powers the human body, fueling every heartbeat, thought, and movement. We'll explore the emerging field of bioelectricity and its role in healing, regeneration, and the potential to unlock new levels of human performance.

Finally, we venture into the mysteries of the mind, where electricity serves as the foundation of consciousness itself. By understanding how electrical signals in the brain create thought and emotion, we open the door to profound insights about what it means to be human. Along the way, we'll connect these scientific discoveries to self-improvement, showing how understanding electricity can help us optimize our lives, boost our well-being, and unlock our potential.

Prepare to journey through the currents of life and consciousness, uncovering how electricity unites the physical and the profound, the ancient and the modern, the scientific and the personal. The story of electricity is, at its heart, the story of life itself.

Part 1: The Birth of Life

Chapter 1: Lightning and the Origin of Life

Early Earth Conditions: A Volatile, Electrically Active Environment

Imagine Earth 4 billion years ago. The planet was young, raw, and brimming with untamed energy. Its skies were heavy with thick clouds of gases—carbon dioxide, methane, ammonia, and water vapor—churning above a landscape of bubbling volcanic activity. Oceans, still forming, spread across the surface, filled with a primordial soup of chemicals waiting for the right conditions to ignite life. It was a world of chaos and potential.

Amid this violent backdrop, lightning storms raged relentlessly. Each bolt of lightning, a flash of raw electrical power, carried enough energy to split molecules apart and create new compounds. Imagine standing on a barren cliff of early Earth—if you could survive it—watching a jagged bolt streak across the sky, connecting heaven and Earth with an electric bridge. This energy didn't just illuminate the sky; it triggered profound chemical reactions that many scientists believe were key to the origin of life.

In the 1950s, scientists Stanley Miller and Harold Urey recreated these early Earth conditions in a laboratory experiment. They combined gases thought to be present in the early atmosphere and exposed them to electrical sparks to simulate lightning. Astonishingly, within days, they found that amino acids—the building blocks of proteins and, ultimately, life itself—had formed. Their groundbreaking work demonstrated how lightning might have provided the spark needed to transform Earth's chemical soup into the precursors of living organisms.

Beyond its role in chemistry, lightning's energy contributed to shaping Earth's environment, helping to break down and recycle materials across the planet's surface. These early processes created a dynamic, electrically active environment where life could take root.

While the conditions of early Earth may seem hostile, they were, in fact, perfect for fostering life. The immense energy from lightning strikes, combined with the rich chemical diversity of the primordial oceans, set the stage for the first molecular steps toward life's emergence. It is humbling to think that something as fierce and chaotic as a lightning storm played such a crucial role in the calm, structured complexity of life we see today.

Understanding these early conditions gives us a deeper appreciation of the forces that shaped us. It reminds us that out of chaos can come order, and from seemingly insurmountable odds, life can emerge. This realization inspires us to look at our own challenges differently—as the raw material for growth, change, and the emergence of something greater.

The Miller-Urey Experiment: How Lightning May Have Sparked Amino Acids and the Building Blocks of Life

In the 1950s, two scientists, Stanley Miller and Harold Urey, set out to answer a profound question: how did life begin on Earth? Inspired by theories suggesting that lightning could have played a role in the origin of life, they decided to replicate the conditions of early Earth in a laboratory. Their experiment became a landmark in understanding how life might have emerged from non-living matter. Miller and Urey started by constructing an apparatus designed to mimic the volatile environment of our planet billions of years ago. They filled a closed system with gases thought to represent Earth's early atmosphere: methane (CH_4), ammonia (NH_3), hydrogen (H_2), and water vapor (H_2O). These gases were chosen because they likely existed in abundance and contained the chemical building blocks for more complex organic molecules.

Next, they introduced a heat source to simulate the warmth of the planet's surface and boiled the water to represent evaporation in the primordial oceans. To simulate lightning, they sent electrical sparks through the gas mixture, mimicking the energy delivered by the violent electrical storms of early Earth. This energy was critical, as it could break molecular bonds and allow new ones to form.

After running the experiment for several days, Miller and Urey made a groundbreaking discovery. Within the liquid that had condensed in their apparatus, they found amino acids—the essential building blocks of proteins. Proteins are vital for life because they perform countless functions within living organisms, from catalyzing chemical reactions as enzymes to forming structural components like hair and muscle.

What made this discovery extraordinary was that amino acids, previously thought to require the presence of living organisms to form, had emerged spontaneously in a simulated early Earth environment. It showed that simple molecules when exposed to the right energy and conditions, could organize themselves into the precursors of life.

The Miller-Urey experiment provided a plausible explanation for how life's foundational chemistry could have arisen naturally. Lightning, a common and powerful source of energy on early Earth, may have acted as the catalyst for these chemical transformations. By repeatedly striking the chemical-rich oceans, lightning could have created a "prebiotic soup" teeming with organic compounds. While the experiment didn't create life itself, it demonstrated that the first steps toward life could occur without the intervention of biology. It bridged the gap between chemistry and biology, showing how non-living matter could give rise to the molecules needed for life.

Since the 1950s, scientists have refined our understanding of early Earth's conditions and repeated similar experiments with different gas mixtures and energy sources. Many of these studies have also yielded amino acids and other organic molecules, supporting the idea that life's building blocks could form naturally in various environments.

Even beyond Earth, the Miller-Urey experiment has implications for astrobiology. It suggests that wherever the right conditions exist—liquid water, a rich chemical environment, and an energy source—there's a possibility for life to emerge. Lightning or other energy sources, such as UV radiation from stars, could spark similar chemical reactions on other planets or moons.

The Miller-Urey experiment not only deepened our understanding of life's origins but also offered a broader message: the potential for transformation lies in the presence of energy and raw materials. Just as lightning turned simple molecules into the precursors of life, challenges and disruptions in our own lives can spark creativity, growth, and new possibilities. At its core, life is the result of harnessing energy and turning chaos into order—a lesson as inspiring as scientific.

The Role of Natural Electric Discharges in Catalyzing Complex Organic Molecules

Natural electric discharges, such as lightning, played a pivotal role in catalyzing the formation of complex organic molecules, which are the precursors to life. In the chaotic environment of early Earth, lightning storms were not rare events but constant and widespread phenomena. These immense electrical discharges carried enough energy to break chemical bonds in simple molecules like methane (CH_4), water vapor (H_2O), ammonia (NH_3), and hydrogen gas (H_2), allowing these atoms to rearrange into more complex compounds.

When lightning strikes the atmosphere, it creates a rapid and extreme increase in temperature, sometimes reaching nearly 30,000 Kelvin—hotter than the surface of the Sun. This intense heat causes molecules to ionize and react with one another, forming unstable intermediates that can recombine into stable organic molecules as the energy dissipates. The nitrogen in ammonia and atmospheric gases, combined with carbon from methane, provided the framework for amino acids and other biologically significant compounds to emerge.

For instance, in the upper layers of the atmosphere, nitrogen oxides created by lightning reactions could dissolve into water droplets, forming nitrate and nitrite compounds. These molecules, when combined with hydrogen and carbon sources in the primordial oceans, could form the basis for nucleotides, the building blocks of RNA and DNA. Similar pathways likely occurred in other chemical families, where electrical energy facilitated the stepwise construction of life-essential molecules.

The energetic process of lightning discharges ensured that these reactions weren't confined to isolated areas. Instead, each bolt provided a localized but potent energy source that could transform atmospheric gases into a rich and diverse chemical soup. Over millions of years, this widespread activity repeatedly delivered essential ingredients into the oceans and created a dynamic, interconnected system of chemical reactions.

Natural electric discharges didn't just stop at forming basic organic molecules. They also catalyzed increasingly complex interactions between these molecules, setting the stage for the emergence of self-replicating systems. Repeated exposure to energy sources like lightning ensured that these organic molecules were not static but continually evolving and combining into larger, more intricate structures. This process of transformation highlights the unique role electricity plays as a catalyst for complexity. Lightning provided the necessary energy to overcome the activation barriers that typically prevent certain chemical reactions from occurring under ordinary conditions. By continuously introducing high-energy inputs, lightning created an environment where simple molecules could form larger, more functional compounds. These reactions weren't limited to amino acids; they likely included sugars, lipids, and other precursors essential for forming the first protocells.

The role of natural electric discharges as a catalyst also speaks to their ability to amplify randomness into order. Each lightning strike introduced a mix of reactions, some leading to dead ends but others creating molecules capable of further interaction and evolution. Over time, this iterative process increased the chemical complexity of the primordial Earth, ultimately contributing to the formation of the first molecular systems capable of self-replication and metabolism.

Understanding this role of electricity provides a window into how life could emerge not only on Earth but also in other electrically dynamic environments across the universe. It demonstrates that electricity, in its raw and powerful form, is not merely a physical phenomenon but a creative force capable of bridging the gap between inanimate matter and the building blocks of life.

Chapter 2: The Earth's Electric Circuit

The Global Electrical System: Earth, Atmosphere, and Ionosphere

The Earth is more than just a rocky planet covered in life; it is part of a vast, interconnected electrical system that extends from its surface to the outer edges of the atmosphere. This system, often called the global electric circuit, operates continuously, linking the ground, the atmosphere, and the ionosphere—a charged layer of the atmosphere around 60 to 1,000 kilometers above the Earth's surface. The global electric circuit is powered by natural electrical processes, including thunderstorms, lightning, and the steady movement of charged particles in the atmosphere. Thunderstorms act as giant electrical generators, transferring positive charges into the upper atmosphere and negative charges to the ground through lightning strikes and other discharges. This ongoing exchange maintains a global voltage gradient, where the ground is negatively charged relative to the atmosphere above it.

The ionosphere plays a critical role in this system. Solar radiation ionizes particles in this layer, creating a vast reservoir of free electrons and positive ions. This ionized layer acts as a conductor, allowing electrical energy generated by thunderstorms and lightning to spread across the planet. Together, the surface of the Earth and the ionosphere form a massive capacitor, separated by the insulating properties of the atmosphere. This capacitor sustains a global electric field, with an average strength of about 100 volts per meter at the Earth's surface under fair-weather conditions.

Thunderstorms and lightning storms, while dramatic and localized, are not the only contributors to this global circuit. Smaller-scale processes, such as the movement of aerosols and the continuous flow of cosmic rays into the atmosphere, also play a role in maintaining this electrical balance. Cosmic rays are high-energy particles that come from outer space. These rays ionize the atmosphere as they collide with air molecules, further influencing the global electric field.

In fair weather, when no storms are present, the global circuit remains active through a phenomenon called fair-weather current. This current, driven by the voltage difference between the ionosphere and the Earth's surface, flows at a rate of about 2 picoamperes per square meter. Though small, this constant current ensures the global electric circuit remains stable and functional, even when storms are absent.

The global electric circuit is not just a scientific curiosity; it has profound implications for life on Earth. It influences weather patterns, cloud formation, and even the movement of charged particles in the atmosphere. Moreover, it provides a framework for understanding how electrical activity connects natural phenomena on a planetary scale. This intricate electrical network demonstrates how interconnected the Earth's systems are, revealing electricity as not just a force of life but a force of the planet itself.

How Atmospheric Electricity Influenced Early Biological Systems

Atmospheric electricity, powered by natural phenomena like lightning and the global electric circuit, played a profound role in shaping early biological systems. Long before life as we know it emerged, this constant flow of electrical energy helped create the conditions necessary for life and influenced the development of early organisms.

Atmospheric electricity provided a steady source of energy capable of driving chemical reactions that would not occur otherwise. In the early Earth environment, lightning and fair-weather electric currents delivered the high-energy inputs required to transform simple gases like methane, ammonia, and carbon dioxide into more complex organic molecules. These molecules, such as amino acids, nucleotides, and lipids, became the building blocks of life.

Electricity's unique ability to alter the structure and properties of molecules also helped stabilize certain reactions, allowing organic compounds to persist in the volatile and dynamic environment of early Earth. Repeated exposure to electrical energy likely fostered the formation of molecular systems capable of storing and transmitting energy, such as adenosine triphosphate (ATP)—a cornerstone of modern biology.

Atmospheric electricity helped establish electrical gradients that may have influenced the structure and behavior of early biological systems. In today's organisms, electrical gradients are essential for processes like nerve signaling, muscle contraction, and cellular energy production. On primordial Earth, similar gradients likely existed in microenvironments, such as near mineral-rich hydrothermal vents or within charged water droplets in the atmosphere.

These natural electrical gradients may have provided the first primitive mechanisms for separating and organizing ions. This process would later become essential for cellular function. For example, the concentration of positively charged ions like sodium and potassium on one side of a barrier could have mimicked the conditions necessary for the first proto-cells to form. These gradients, maintained by atmospheric and environmental electricity, may have driven the evolution of energy storage and transfer systems.

As life began to emerge, atmospheric electricity likely continued to play a role in stimulating early biological systems. Primitive cells, surrounded by membranes sensitive to electrical charges, could have responded to changes in the global electric field or localized discharges, influencing their movement, replication, or interaction with their environment.

For example, the constant electrical stimulation from fair-weather currents and lightning strikes may have spurred chemical changes within early life forms, accelerating their ability to adapt and evolve. These electrical interactions could also have enhanced the development of bioelectric communication systems—precursors to the complex signaling pathways we see in modern organisms.

The dynamic nature of atmospheric electricity provided an ever-changing environment that likely drove evolutionary innovation. Electrical discharges introduced variability into chemical systems, creating new molecules and structures that early life forms could exploit. Over time, organisms that could harness or withstand atmospheric electrical forces may have gained an evolutionary advantage, leading to the development of bioelectric systems that are now fundamental to life.

By influencing chemical reactions, creating gradients, and stimulating biological activity, atmospheric electricity laid the groundwork for life to emerge and evolve. It not only helped spark the first organic molecules but also shaped the environment in which early biological systems developed, highlighting electricity's integral role in the history of life.

Potential Links Between Earth's Natural Electric Fields and Early Microbial Evolution

Earth's natural electric fields, generated by the global electric circuit and localized electrical phenomena, likely played a significant role in shaping the evolution of early microbes. These fields created a unique environment where electrical forces interacted with primitive life forms, influencing their survival, adaptation, and eventual complexity.

One of the earliest potential interactions between microbes and Earth's natural electric fields is electrotaxis, the ability of organisms to move in response to electrical gradients. Modern bacteria, such as *Escherichia coli* and *Pseudomonas aeruginosa*, exhibit electrotaxis, suggesting that this behavior may have ancient roots.

On early Earth, natural electric fields in environments like hydrothermal vents, volcanic areas, or even within water droplets charged by lightning could have guided microbial movement. By responding to electrical gradients, early microbes may have been directed toward energy-rich environments or areas with essential nutrients. This ability would have given them a survival advantage, promoting the development of electrosensitive systems over time.

Natural electric fields provided primitive organisms with a potential energy source. Early microbes may have evolved mechanisms to exploit these fields, much like modern organisms utilize bioelectric gradients across their membranes for energy production. For instance:

- **Proton Gradients:** Natural electric fields could have driven the movement of charged particles, such as protons, across primitive membranes. This process resembles the proton gradients used in modern cellular respiration and photosynthesis to generate ATP, the universal energy currency of life.
- **Electron Transfer Chains:** Microbes that evolved the ability to transfer electrons across membranes might have harnessed Earth's electric fields to drive energy-producing reactions. This capability is seen today in some bacteria that can metabolize minerals, such as *Geobacter* and *Shewanella*, which use external electron acceptors in their environment.

The interaction between natural electric fields and primitive membranes may have accelerated the evolution of membrane structures. Early cell membranes were likely simple lipid bilayers, but exposure to electrical forces could have caused changes in their permeability, stability, and ability to maintain ion gradients. Electric fields may have also encouraged the formation of voltage-sensitive channels, proteins that regulate ion flow in response to electrical changes. These channels are critical in modern cells for maintaining electrical potential and signaling, suggesting that Earth's natural electric fields could have driven their early development.

The constant exposure of early microbes to fluctuating electric fields could have introduced selective pressures that shaped their evolution. Organisms that could tolerate or even exploit these fields would have had a distinct advantage, leading to the development of bioelectric capabilities.

Additionally, electrical forces may have influenced genetic processes. For example:

- Electric fields could have affected the folding and stability of early nucleic acids, promoting the evolution of more stable and functional genetic material.
- Electrical interactions may have played a role in the assembly and replication of RNA and DNA, critical steps in the evolution of life.

Earth's electric fields could have influenced the formation of early microbial communities, such as biofilms. Biofilms are clusters of microorganisms that stick to surfaces and communicate through chemical and electrical signals. In modern systems, biofilms often form near charged surfaces or in environments with strong electrical gradients. This suggests that early microbes may have used natural electric fields to aggregate and share resources.

These communal structures would have allowed microbes to exchange genetic material and cooperate, increasing their adaptability and resilience in harsh early Earth environments. Over time, such interactions likely drove the evolution of more complex and cooperative behaviors.

Natural electric fields provided a dynamic and challenging environment that forced early microbes to innovate. By interacting with these fields, microbes likely developed mechanisms that laid the foundation for bioelectricity in more complex organisms. The ability to sense, respond to, and harness electrical forces became an essential evolutionary tool, shaping the trajectory of life from its earliest stages.

Earth's natural electric fields, therefore, were not merely passive background phenomena but active agents in driving microbial evolution. They connected the physical and biological realms, providing both challenges and opportunities that early life forms used to thrive and evolve. This interaction between microbes and electric fields underscores the profound influence of Earth's electrical environment on the emergence and diversification of life.

Part 2: Electricity and Evolution

Chapter 3: Electricity in Plant Life

How Plants Generate and Use Bioelectric Signals for Communication, Growth, and Defense

Though seemingly passive, plants are dynamic electrical systems that utilize bioelectric signals to interact with their environment, regulate their growth, and defend themselves. While plants lack the nervous systems seen in animals, they possess sophisticated electrical networks that enable them to communicate internally and respond to external stimuli in remarkable ways.

At the heart of plant bioelectricity lies the movement of ions, such as potassium, calcium, and chloride, across cell membranes. This ionic flow generates electrical potentials, creating a charged environment that plants use to send signals throughout their tissues. One of the most well-known examples of this is the Venus flytrap, a carnivorous plant that uses electrical signals to snap its trap shut. When an insect touches the sensory hairs inside the trap, it triggers a rapid shift in ion concentrations, creating an electrical signal that travels across the plant cells and causes the trap to close.

Beyond individual responses, plants also use electrical signaling to coordinate growth and development. Electrical potentials help direct resources to areas of the plant that require nutrients or repair. For instance, when a leaf is damaged, bioelectric signals travel from the wound site to other parts of the plant, triggering the production of protective chemicals. These signals ensure that the plant can respond efficiently to injuries, sealing the wound and strengthening nearby tissues.

Plants also use bioelectricity to sense and adapt to their surroundings. Roots, for example, generate electrical gradients that help them navigate the soil in search of water and nutrients. These gradients allow the roots to respond to environmental cues, such as the presence of minerals or changes in moisture levels, ensuring the plant can sustain itself even in challenging conditions.

Defense is another critical area where plant bioelectricity plays a role. When under attack by herbivores or pathogens, plants emit electrical signals that act as distress calls, activating defensive mechanisms throughout the plant. These signals can lead to the release of chemicals that deter pests or the production of proteins that combat infections. Additionally, some plants use electrical signals to communicate with neighboring plants, warning them of potential threats and prompting them to prepare their defenses.

Plant bioelectricity operates on a slower scale than the nervous systems of animals. Yet, it is no less intricate or vital. By generating and interpreting electrical signals, plants demonstrate a level of adaptability and resilience that belies their stationary nature. These electrical processes are a testament to the ingenuity of evolution, showcasing how life can harness fundamental forces like electricity to thrive in diverse and dynamic environments.

The study of plant bioelectricity not only deepens our understanding of how plants function but also inspires new ways to approach technology and sustainability. Researchers are exploring how plant-generated electricity could be harnessed as a renewable energy source or used to develop bio-inspired systems for environmental monitoring. By learning from the silent but powerful electrical lives of plants, we gain insights that could transform how we interact with the natural world and design solutions for the future.

Electrical Signaling in Roots and Leaves: the "Nervous System" of Plants

Plants may not have brains or nerves, but they possess a remarkable ability to communicate internally through electrical signaling, creating a system akin to a rudimentary nervous system. This network of electrical activity, facilitated by roots and leaves, allows plants to sense their environment, relay information, and coordinate responses throughout their structure.

At the cellular level, electrical signaling in plants is driven by the movement of ions across membranes, generating electrical potentials. This process, known as action potential, is similar in principle to how neurons in animals transmit signals. When a stimulus, such as a touch, injury, or change in light, is detected, ion channels in plant cells open, allowing ions like calcium, potassium, and chloride to flow in and out of cells. This creates an electrical wave that propagates across the plant, transmitting information from one part to another.

The roots of a plant are a hub of electrical activity, constantly gathering information about the soil's conditions. Through electrical gradients, roots detect variations in moisture, nutrients, and the presence of harmful substances. These signals are then relayed to other parts of the plant, prompting adjustments in water uptake, nutrient distribution, or root growth direction. For example, when a section of roots encounters dry soil, electrical signals trigger the closure of stomata—tiny openings on the leaves—to conserve water.

Leaves, on the other hand, are crucial for sensing external stimuli such as light, temperature, and mechanical disturbances. When a leaf is damaged by a herbivore or an environmental stressor, it generates an electrical signal that travels through the plant, warning other leaves of the attack. This rapid communication enables the plant to activate defensive mechanisms, such as producing bitter compounds to deter herbivores or enzymes that combat infections. In this way, electrical signaling ensures that a local injury can elicit a whole-plant response.

Interestingly, the electrical signals in plants are often accompanied by chemical signals, creating a robust communication system. For instance, calcium ions act as messengers that amplify the electrical signals, ensuring they reach their target efficiently. These combined electrical and chemical responses are what make plant signaling so versatile and adaptive despite the absence of traditional nerves.

The term "nervous system" for plants is metaphorical, but it underscores the sophistication of these mechanisms. Plants exhibit a level of internal communication and environmental awareness that challenges traditional views of their simplicity. Their ability to generate, propagate, and respond to electrical signals allows them to survive and thrive in ever-changing environments.

The study of this plant "nervous system" not only reveals the hidden complexities of plant biology but also offers insights into bio-inspired technologies. Understanding how plants transmit and respond to electrical signals could inspire advancements in agriculture, such as developing crops that can better adapt to stress, or innovations in engineering, such as plant-based sensors for environmental monitoring. Through their silent, electrical dialogue, plants remind us of the remarkable ways life harnesses fundamental forces to solve challenges and sustain itself.

The Role of Plant Bioelectricity in Shaping Ecosystems

Plant bioelectricity plays a subtle but profound role in shaping ecosystems, creating interconnected networks of communication and influence that sustain life across entire landscapes. Through their electrical signaling systems, plants not only manage their own survival but also interact with other organisms and the environment, contributing to the overall health and balance of ecosystems.

One of the most critical ways plant bioelectricity impacts ecosystems is through the regulation of nutrient cycles. Roots generate electrical gradients that guide their growth and activity, enabling plants to extract essential nutrients like nitrogen, phosphorus, and potassium from the soil. As plants absorb these nutrients and incorporate them into their structures, they eventually return them to the soil through leaf litter and decay, creating a cycle that supports microbial communities and other plants. Electrical signals within roots can also influence their symbiotic relationships with fungi, such as mycorrhizae, which help plants access nutrients in exchange for carbon compounds. These symbiotic partnerships are foundational to nutrient distribution in ecosystems.

Plant bioelectricity also plays a significant role in water management within ecosystems. By regulating the opening and closing of stomata, the tiny pores on leaf surfaces, plants control water loss through transpiration. This process not only sustains the plant but also contributes to the movement of water through the environment, supporting local humidity levels and influencing weather patterns. Electrical signals triggered by drought stress can reduce transpiration, conserving water in times of scarcity and maintaining stability in the ecosystem's water cycle.

Plants use bioelectric signals to communicate with one another and even with other species, affecting ecosystem dynamics. For example, when a plant experiences herbivory or mechanical damage, it sends electrical signals throughout its tissues, prompting the release of volatile organic compounds (VOCs) into the air. These VOCs can alert neighboring plants, triggering preemptive defensive responses such as the production of protective chemicals. This form of plant-to-plant communication creates a network of mutual awareness, enhancing the resilience of plant communities.

Beyond their individual survival, plants also use electrical activity to maintain symbiotic relationships with other organisms. Electrical gradients in roots help facilitate partnerships with nitrogen-fixing bacteria. This bacteria converts atmospheric nitrogen into a form that plants can use. These bacteria, in turn, enrich the soil, supporting other plants and sustaining the broader ecosystem. Similarly, plants with bioelectric responses to environmental cues can provide critical resources, such as nectar or fruit, at optimal times for pollinators or seed dispersers, fostering biodiversity and ensuring the continuation of life cycles.

Plant bioelectricity influences ecosystem stability and resilience, especially in the face of disturbances like climate change or habitat destruction. Electrical signaling allows plants to respond dynamically to environmental stressors, such as temperature shifts or soil degradation, and adapt to changing conditions. As keystone species in many ecosystems, their ability to maintain balance through these adaptive responses supports entire communities of organisms.

The role of plant bioelectricity in shaping ecosystems highlights the interconnectedness of life. Plants act as silent mediators, transmitting and responding to electrical signals that affect nutrient flow, water cycles, and interspecies interactions. By understanding these processes, scientists can better appreciate the intricate web of relationships that sustain ecosystems and explore new ways to restore degraded environments or enhance agricultural sustainability. Plant bioelectricity is not merely a feature of individual organisms; it is a vital force that contributes to the harmony and function of life on Earth.

Chapter 4: Bioelectricity in Animal Evolution

Early Multicellular Organisms and the Rise of Bioelectric Communication

The emergence of multicellular organisms marked a transformative period in the history of life on Earth, and bioelectric communication played a pivotal role in this evolution. Single-celled organisms, such as bacteria, were already utilizing electrical signals to communicate and coordinate behaviors within colonies. However, as life evolved into more complex multicellular forms, the need for efficient communication between cells within a single organism became essential. This was where bioelectricity provided a significant advantage.

In early multicellular organisms, electrical signals allowed cells to share information rapidly, enabling them to coordinate activities like movement, feeding, and defense. Ion channels, which regulate the flow of charged particles across cell membranes, became the foundation of this communication system. These channels allowed for the generation and propagation of electrical signals, creating a rudimentary network for intercellular communication. For example, electrical gradients may have helped early multicellular organisms contract their primitive muscle-like structures, enabling movement that improved their ability to find food or escape predators.

The simplicity of bioelectric communication allowed these early organisms to adapt to environmental changes quickly. This adaptability, coupled with the efficiency of electrical signaling compared to slower chemical communication, laid the groundwork for the evolution of more sophisticated systems. As multicellular organisms became larger and more complex, the ability to transmit electrical signals over long distances within the body became increasingly important, setting the stage for the evolution of nervous systems.

The Evolution of Nervous Systems: Why Electricity Became Central to Animal Function

The evolution of nervous systems represented a revolutionary leap in animal complexity, and electricity was at the core of this advancement. Nervous systems arose as specialized networks of cells capable of generating, transmitting, and processing electrical signals. This allowed animals to interact with their environment in ways that were far more dynamic and responsive than ever before.

Nervous systems began with simple nerve nets, as seen in modern-day cnidarians like jellyfish. These networks lacked a centralized brain but allowed electrical signals to travel between interconnected neurons, coordinating basic movements such as pulsing or tentacle contraction. Even at this primitive stage, the speed of electrical transmission provided a critical survival advantage, enabling rapid responses to threats or opportunities.

As animals continued to evolve, nervous systems became more centralized and complex. Ganglia—clusters of nerve cells—emerged in certain species, allowing for localized processing of electrical signals. Eventually, the central nervous system, including the brain and spinal cord, appeared in more advanced animals.

This centralization allowed for the integration of sensory input, decision-making, and coordination of motor functions, giving rise to behaviors that were faster, more precise, and more adaptive.

Electricity became central to animal function because it provided unparalleled speed and efficiency. Electrical signals travel nearly instantaneously along neurons, far outpacing chemical signaling. This made electricity indispensable for animals with complex structures, enabling quick reflexes, precise coordination of muscles, and the integration of sensory information into cohesive responses. Over time, the reliance on electrical signaling became a hallmark of animal life, underpinning critical functions such as hunting, mating, and avoiding predators.

Specialized Creatures: Electric Fish and the Evolutionary Advantage of Electricity

The evolutionary story of bioelectricity reaches a fascinating pinnacle in specialized creatures like electric fish, which have harnessed electricity not just for communication or movement but as a powerful tool for survival. Electric fish, such as electric eels, electric rays, and some species of catfish, have evolved specialized cells called electrocytes, which generate and store electrical energy. These cells function much like biological batteries, creating electrical discharges used for hunting, defense, and navigation.

Electric eels, for example, can produce electric shocks of up to 600 volts, enough to stun or kill prey and deter predators. By discharging this energy, they can immobilize fish or small animals, making hunting far more efficient. Similarly, electric rays use their discharges defensively, delivering shocks to anything that threatens them.

In addition to offense and defense, many electric fish use weaker electrical signals for navigation and communication. These signals create an electric field around the fish, which interacts with objects and other organisms in their environment. By detecting distortions in this field, the fish can locate prey, avoid obstacles, and even identify other individuals of their species. This ability, known as electrolocation, provides a significant advantage in murky waters where visibility is limited.

The evolutionary innovation of electric organs demonstrates how bioelectricity can be adapted for highly specialized functions. These adaptations underscore the versatility of electricity as a biological tool and highlight its centrality to life's evolutionary story. Electric fish, in particular, are a vivid example of how nature continually refines and reimagines biological systems to meet the challenges of survival and reproduction.

Bioelectricity has been an essential driver of animal evolution, from its origins in primitive communication systems to its role in the development of advanced nervous systems and specialized adaptations like electric organs. By harnessing the power of electricity, animals have achieved remarkable levels of complexity, adaptability, and success in navigating the challenges of life. This evolutionary journey underscores electricity's fundamental importance in shaping the diversity of life on Earth.

Chapter 5: The Role of Electricity in Human Evolution

How Bioelectricity Influenced Brain Development and Complexity

Bioelectricity has been a crucial factor in the evolution of the human brain, which is often regarded as the most complex biological structure on Earth. The brain operates entirely on electrical signals, using neurons to transmit information across vast networks. The evolution of this intricate bioelectric system allowed early humans to develop advanced cognitive abilities, problem-solving skills, and complex social behaviors.

The early stages of brain evolution in vertebrates were marked by the expansion of neural networks, made possible by the efficiency of electrical communication. As neural circuits became denser and more interconnected, the brain's capacity to process information increased. This allowed for better coordination of movement, heightened sensory perception, and, eventually, more sophisticated learning and memory systems.

In humans, the evolution of the cerebral cortex—a region of the brain responsible for higher-order functions like reasoning, language, and abstract thought—was made possible by innovations in bioelectricity. The ability to rapidly transmit and integrate electrical signals across vast neural networks allowed the human brain to develop unparalleled complexity. This complexity not only enabled humans to adapt to their environment but also facilitated the creation of tools, language, and culture, all of which became defining features of humanity.

The Interplay of Diet, Bioelectricity, and Energy Demands in Humans

The evolution of the human brain was not solely a matter of increasing complexity in bioelectric signaling; it also depended on meeting the brain's immense energy demands. The human brain consumes about 20% of the body's total energy despite accounting for only 2% of its weight. This energy is used to maintain the brain's bioelectric activity, including the generation of action potentials, the release of neurotransmitters, and the upkeep of ion gradients across cell membranes.

Diet played a pivotal role in supporting the energy needs of this bioelectric system. The shift to a diet rich in nutrient-dense foods, particularly those high in fats and proteins, provided the caloric surplus needed to sustain a growing and energy-intensive brain. Access to these foods may have been facilitated by technological advancements, such as cooking and tool use, which allowed early humans to process and extract more energy from their environment.

Specific nutrients like omega-3 fatty acids, found in fish and other sources, are essential for maintaining the structure and function of neural membranes, which are critical for electrical signaling. The availability of these nutrients likely influenced the evolutionary trajectory of the human brain, enabling it to maintain the bioelectric processes necessary for advanced cognitive functions.

This interplay between diet, energy demands, and bioelectricity highlights the delicate balance that supported human brain evolution. Without the ability to sustain the electrical activity that drives neural function, the remarkable cognitive abilities of humans would not have been possible.

Possible Connections Between Electricity and the Evolution of Human Consciousness

The evolution of human consciousness is one of the most profound and mysterious aspects of our history, and bioelectricity may hold some clues to its origins. Consciousness, as experienced by humans, arises from the coordinated activity of billions of neurons, all communicating through electrical impulses. This dynamic electrical activity creates brainwaves, which correspond to different states of awareness, from deep sleep to focused attention.

One hypothesis suggests that the increasing complexity of electrical signaling in the human brain, coupled with its capacity for integration across diverse neural regions, led to the emergence of self-awareness and higher consciousness. The synchronization of electrical activity in various parts of the brain, known as neural coherence, is thought to be critical for unifying sensory input, memory, and thought into a cohesive experience.

The electrical rhythms of the brain also play a role in emotional states, decision-making, and creativity, all of which are hallmarks of human consciousness. For example, the prefrontal cortex, a region associated with planning and self-control, relies on precise electrical coordination to function effectively. These bioelectric processes allowed early humans to navigate complex social structures, anticipate future events, and innovate in ways that distinguished them from other species.

Another intriguing possibility is that bioelectricity in the human brain interacts with environmental electric fields. While speculative, some researchers have proposed that Earth's natural electric fields or electromagnetic phenomena might subtly influence brain activity, contributing to the evolution of consciousness. This idea aligns with the broader concept of humans as part of an interconnected electrical system, both internally and externally.

The role of electricity in the evolution of human consciousness underscores the deep connection between biology and energy. Bioelectricity enabled the human brain to not only grow in size and complexity but also develop the capacity for introspection, creativity, and cultural expression. By understanding these processes, we gain insight into what it means to be human and how the forces of nature shaped the mind's extraordinary potential.

Part 3: Electricity and the Human Body

Chapter 6: How Electricity Powers Life

Recap of Human Bioelectricity: Cells, Tissues, and Organ Systems

Human bioelectricity is the foundation of life, enabling every cell, tissue, and organ in the body to function. At its core, bioelectricity is generated by the movement of charged particles, or ions, across cell membranes. This movement creates electrical potentials that allow cells to communicate, coordinate activities, and respond to their environment.

Each cell in the human body is surrounded by a membrane that separates its internal and external environments. Embedded within this membrane are ion channels and pumps that regulate the flow of sodium, potassium, calcium, and chloride ions. These charged particles create electrical gradients, with the inside of the cell typically more negatively charged than the outside. This difference in charge is known as the resting membrane potential, a critical element of cellular bioelectricity.

Beyond individual cells, tissues and organ systems rely on electrical signals to maintain coordinated function. For example, clusters of specialized cells, such as neurons in the brain or pacemaker cells in the heart, generate and propagate electrical signals that control everything from muscle contractions to thought processes. This network of bioelectric communication is the basis of human physiology, ensuring that the body's systems work together seamlessly.

The Role of Ion Channels and Action Potentials in the Nervous System

The nervous system is a masterful conductor of bioelectricity, using electrical signals to transmit information rapidly across the body. At the heart of this system are neurons, the specialized cells responsible for generating and propagating action potentials, which are rapid changes in electrical charge across the cell membrane. Ion channels, which are protein structures embedded in the membrane, play a central role in this process. When a neuron is activated by a stimulus, ion channels open to allow sodium ions to rush into the cell, temporarily reversing the charge across the membrane. This creates a wave of electrical activity that travels along the length of the neuron. As the signal propagates, potassium ions exit the cell to restore the resting membrane potential, resetting the neuron for the next signal. These action potentials allow neurons to communicate with one another across synapses, the small gaps between cells. At the synapse, the electrical signal triggers the release of neurotransmitters, chemical messengers that carry the signal to the next neuron or target cell. This intricate process is the foundation of everything the nervous system does, from sensing pain to forming memories.

The precision and speed of electrical signaling in the nervous system make it one of the most efficient communication networks in nature. Without ion channels and the bioelectric phenomena they enable, the complex coordination of bodily functions and cognitive processes would be impossible.

Electrical Activity in the Heart, Muscles, and Other Tissues

Electrical activity is not confined to the nervous system; it is vital for the function of the heart, muscles, and various other tissues. The heart, for example, relies on a specialized electrical system to maintain its rhythmic contractions. Pacemaker cells in the sinoatrial node generate electrical impulses that travel through the heart muscle, triggering coordinated contractions that pump blood throughout the body. Electrocardiograms (ECGs) measure this electrical activity, providing a window into the heart's health and rhythm.

In skeletal muscles, electrical signals originating in the nervous system stimulate contraction. When a motor neuron sends an action potential to a muscle fiber, it triggers the release of calcium ions within the muscle cell. This calcium influx activates the proteins responsible for muscle contraction, enabling movement. The process is highly energy-dependent, requiring precise coordination of electrical and chemical signals.

Even non-excitable tissues, such as epithelial cells, exhibit electrical properties that are essential for their function. For example, electrical gradients drive the movement of ions and fluids across cell membranes, regulating processes like nutrient absorption in the intestines or the production of sweat in glands.

Electrical activity also plays a role in wound healing and tissue regeneration. When a tissue is injured, cells at the site generate bioelectric signals that attract immune cells and promote the repair process. These signals help guide the formation of new tissue, ensuring that the body can recover effectively from damage.

In every part of the body, electricity powers life by coordinating the activities of cells, tissues, and organ systems. From the rhythmic beating of the heart to the intricate firing of neurons, bioelectricity is the invisible force that sustains us. Understanding how it works not only deepens our appreciation for the body's complexity but also opens the door to new ways of enhancing health and well-being.

Chapter 7: Healing and Regeneration

Natural Electric Fields in Wound Healing and Tissue Repair

The human body has remarkable self-healing mechanisms, and natural electric fields play a vital role in these processes. When an injury occurs, the body generates electrical signals around the wound, known as injury currents or endogenous electric fields. These fields are created by the flow of ions through the damaged tissue and serve as a bioelectric roadmap for the cells responsible for healing.

At the site of a wound, epithelial cells and fibroblasts, which are critical for tissue repair, migrate toward the area with the highest electrical charge. This process, known as electrotaxis, ensures that cells move in a coordinated manner to close the wound, rebuild the tissue, and restore function. Electric fields also attract immune cells, such as macrophages, to the site of injury, helping to clear debris and fight infection.

The presence of natural electric fields accelerates processes like cell proliferation, tissue regeneration, and angiogenesis—the formation of new blood vessels. For example, when a skin wound begins to close, electrical signals help guide the cells forming the new epidermis, ensuring that the skin heals seamlessly. These bioelectric signals also play a role in nerve regeneration, stimulating the growth of damaged nerve fibers to restore sensation and mobility.

Understanding the role of natural electric fields in healing provides practical insights into improving recovery times. By maintaining optimal nutrition and hydration, individuals can support the ionic balance necessary for efficient electrical signaling during tissue repair. Additionally, reducing stress, which can disrupt bioelectric activity, may enhance the body's natural ability to heal.

Advances in Bioelectric Medicine: Electroceuticals and Regeneration

The understanding of bioelectricity in healing has led to groundbreaking advances in medicine, particularly in the field of electroceuticals. Electroceuticals are medical devices or therapies that use electrical stimulation to treat injuries and illnesses, offering a non-invasive and often drug-free alternative to traditional treatments.

One of the most well-established applications of bioelectric medicine is in bone healing. Devices that deliver low-intensity electrical stimulation are used to treat non-healing fractures, helping to promote bone growth and repair. These devices mimic the natural electric fields generated during bone remodeling, encouraging the activity of osteoblasts—the cells responsible for forming new bone.

Electrical stimulation has also shown promise in treating chronic wounds, such as those caused by diabetes or pressure ulcers. Clinical studies have demonstrated that applying electric fields to these wounds can significantly accelerate healing by enhancing cell migration and increasing blood flow to the affected area.

Electroceuticals are also being explored in nerve repair and spinal cord injuries. Devices that deliver targeted electrical impulses to damaged nerves can stimulate regrowth, improve function, and reduce pain. For example, implantable stimulators have been used to restore movement in individuals with paralysis by activating specific nerve pathways.

Beyond repairing damage, bioelectric medicine is venturing into the realm of regeneration. Researchers are developing techniques to use electrical signals to encourage the regrowth of tissues and even entire organs. For instance, experiments with amphibians, such as frogs and salamanders—known for their natural regenerative abilities—have revealed that bioelectric cues play a critical role in limb regeneration. By mimicking these signals in mammals, scientists hope to unlock similar regenerative potential in humans.

These advances in bioelectric medicine have practical implications for improving quality of life. Whether it's accelerating recovery from an injury, managing chronic pain, or potentially regrowing lost tissue, electroceuticals offer hope for more effective and personalized healthcare solutions.

Speculative Theories about Limb and Organ Regeneration

The idea of regenerating lost limbs or damaged organs has fascinated humans for centuries, and bioelectricity may hold the key to turning this dream into reality. While mammals have limited regenerative capabilities, certain species, like salamanders, can regrow entire limbs. Research into these organisms has revealed that bioelectric signals are integral to this process, serving as the blueprint for tissue regeneration.

In salamanders, the loss of a limb triggers a cascade of bioelectric activity at the injury site. Ion channels and pumps create specific electrical patterns that guide stem cells to the area where they proliferate and differentiate into the necessary cell types for regrowth. These bioelectric signals also influence the spatial organization of tissues, ensuring that the new limb develops with the correct structure and functionality.

Scientists are now exploring ways to replicate these processes in humans. One speculative approach involves reprogramming human cells using bioelectric cues. By altering the electrical environment of a wound site, researchers hope to stimulate the activation of dormant stem cells and direct their development into new tissues. Early experiments have shown promise, with some studies successfully regenerating partial limb structures in animal models.

Another avenue of research focuses on organ regeneration. Electrical stimulation has been used to encourage the growth of functional tissue in damaged organs, such as the heart and liver. For example, bioengineered scaffolds that mimic the natural electric fields of these organs have been shown to improve cell growth and organization, paving the way for the creation of fully functional replacement organs.

While these theories remain in the experimental stage, the potential applications are vast. Imagine a future where amputees could regrow their limbs or individuals with organ failure could regenerate their own tissues without the need for transplants. Such breakthroughs would revolutionize medicine and redefine what is possible for human healing.

Beyond the scientific implications, the concept of regeneration carries a powerful message for personal growth and resilience. Just as bioelectricity enables cells and tissues to rebuild after injury, humans possess the capacity to recover and rebuild in the face of challenges. Understanding and harnessing the principles of regeneration can inspire a mindset of renewal, encouraging individuals to see obstacles as opportunities for growth.

Bioelectricity's role in healing and regeneration is not just a scientific curiosity; it is a testament to the body's innate potential to restore and adapt. By unlocking the secrets of this natural force, we may one day achieve unprecedented advancements in medicine and transform the way we approach health and recovery.

Chapter 8: The Schumann Resonance and the Human Body

Earth's Natural Resonances and their Potential Connection to Human Biology

The Schumann Resonance is a natural electromagnetic phenomenon that occurs in the space between the Earth's surface and its ionosphere. It is created by lightning strikes, which generate low-frequency electromagnetic waves that resonate within this cavity. These resonances are remarkably consistent, with a fundamental frequency of approximately 7.83 Hz and several harmonics. Often referred to as Earth's "heartbeat," the Schumann Resonance is a subtle but pervasive presence in our environment, and researchers have speculated about its potential effects on human biology.

Human beings are immersed in this electromagnetic environment, and some scientists suggest that the Schumann Resonance might interact with our own bioelectric systems. The human brain operates on electrical impulses, generating waves in various frequency ranges, such as delta, theta, alpha, beta, and gamma waves. Interestingly, the alpha brainwave range, which is associated with relaxation and calm focus, overlaps with the fundamental frequency of the Schumann Resonance. This overlap has sparked interest in understanding whether Earth's natural electromagnetic rhythms could influence brain activity and other biological processes.

The idea that life on Earth has evolved in harmony with the planet's electromagnetic environment is not far-fetched. Just as plants and animals are influenced by natural cycles like sunlight and seasons, some researchers believe that Earth's electromagnetic field might serve as a subtle regulator for certain biological functions. While this concept is still under investigation, it opens the door to exploring how humans are connected to the planet at an energetic level.

Theories Linking the Schumann Resonance to Brain Activity and Circadian Rhythms

One of the most intriguing hypotheses about the Schumann Resonance is its potential connection to brain activity. The human brain's electrical signals create oscillations, or brainwaves, that govern states of consciousness, mood, and cognitive function. Alpha waves, which operate at frequencies between 8 and 12 Hz, are particularly interesting because they fall within the range of the Schumann Resonance's harmonics. These waves are typically associated with meditative states, creativity, and relaxation.

Some researchers propose that the Schumann Resonance could act as an external synchronizing force for these alpha brainwaves, promoting mental clarity and emotional balance. Anecdotal reports suggest that exposure to environments rich in natural electromagnetic frequencies—such as forests, mountains, or beaches—can have calming and restorative effects, potentially due to alignment with the Schumann Resonance. While definitive scientific evidence is lacking, these observations hint at a possible connection between Earth's natural rhythms and human well-being.

The Schumann Resonance has also been linked to circadian rhythms. This internal clock governs sleep-wake cycles and other physiological processes. Circadian rhythms are regulated by external cues such as light and temperature. Still, some theories suggest that electromagnetic signals, including the Schumann Resonance, play a secondary role. Proponents of this idea argue that the stability of these natural frequencies could help the body maintain its internal clock, particularly in the absence of light-based cues.

In a modern world filled with artificial electromagnetic noise from electronic devices, power lines, and wireless communications, it's possible that human bioelectric systems may become desynchronized from natural rhythms. Advocates for reconnecting with the Schumann Resonance suggest that spending time in nature or reducing exposure to artificial electromagnetic fields might help restore harmony and improve overall health.

Separating Science from Speculation

While the Schumann Resonance is a well-documented physical phenomenon, its effects on human biology remain speculative. There is little direct scientific evidence linking the Schumann Resonance to brain activity, circadian rhythms, or overall health. Much of the research in this area is theoretical or anecdotal, with few controlled studies to confirm the proposed effects.

One challenge in studying this topic is the subtlety of the Schumann Resonance. The electromagnetic waves it produces are extremely weak compared to the artificial electromagnetic fields generated by modern technology. Measuring their influence on human biology requires highly sensitive equipment and carefully controlled environments, making it difficult to draw definitive conclusions.

Critics argue that the perceived effects of the Schumann Resonance may be psychological rather than physiological. The calming sensation associated with natural environments, for example, could be attributed to factors like reduced stress, fresh air, and the absence of urban noise rather than electromagnetic synchronization. Without rigorous experimental data, it is impossible to determine whether the Schumann Resonance has a direct biological impact or if it is simply one aspect of a broader environmental experience.

However, the lack of definitive proof does not entirely discount the potential significance of the Schumann Resonance. The overlap between its frequencies and human brainwave patterns is intriguing, and further research could reveal subtle interactions that contribute to our understanding of bioelectricity and human health. Regardless of its direct effects, the concept of aligning with Earth's natural rhythms encourages mindfulness of our connection to the planet and the ways our environment influences well-being.

By separating science from speculation, we can appreciate the Schumann Resonance as a fascinating natural phenomenon while continuing to explore its potential implications for human biology. Whether as a scientific inquiry or a metaphor for harmony with the Earth, it reminds us of the intricate and unseen connections that bind life to the planet.

Part 4: Electricity and Consciousness

Chapter 9: Brainwaves and the Electric Mind

How the Brain Generates and Uses Electrical Signals to Create Thought and Emotion

The human brain is an electrical marvel, a dense network of around 86 billion neurons communicating with each other to produce thought, emotion, and behavior. At the core of this activity is bioelectricity, the movement of charged particles called ions across neuronal membranes. This process generates tiny electrical impulses, known as action potentials, that travel along neurons to transmit information.

Neurons use specialized proteins called ion channels to regulate the flow of sodium, potassium, and calcium ions. When a neuron is activated by a stimulus—whether external, like the sight of a sunset, or internal, like a memory—a rapid exchange of ions occurs, creating an electrical charge that propagates down the neuron's axon. At the end of the axon, this electrical signal triggers the release of chemical messengers called neurotransmitters, which carry the signal to neighboring neurons.

This bioelectric communication is extraordinarily efficient, enabling the brain to process vast amounts of information in milliseconds. Electrical signals underlie every aspect of brain function, from basic survival instincts to the creation of art and the experience of love. Thought and emotion emerge from complex patterns of electrical activity across interconnected networks of neurons, giving rise to the richness of human consciousness.

The Relationship Between Brainwave Frequencies and States of Consciousness

The brain's electrical activity is not random; it produces rhythmic patterns of oscillations known as brainwaves. These brainwaves are categorized by their frequencies, measured in hertz (cycles per second). They are associated with different states of consciousness and mental activity.

Delta waves, the slowest brainwaves, dominate during deep sleep and are crucial for physical and emotional restoration. Theta waves, slightly faster, are linked to light sleep, relaxation, and creative insights. These waves often occur during moments of daydreaming or meditation when the mind is in a state of quiet reflection.

Alpha waves, operating at 8–12 Hz, are associated with a relaxed yet alert state. They are often observed when individuals are calm, focused, or engaged in creative activities. Beta waves, which are faster, are linked to active thinking, problem-solving, and attention. Finally, gamma waves, the fastest brainwaves, are associated with high-level cognitive functioning, such as learning, memory, and sensory integration.

The transitions between these states of consciousness are governed by shifts in brainwave activity, which can be influenced by external stimuli, internal thoughts, or deliberate practices like meditation and mindfulness. Understanding these brainwave frequencies provides insights into how humans experience the world and how they can actively shape their mental states.

For example, practices like deep breathing or meditation can help slow brainwave activity from the faster beta state to the more relaxed alpha or theta states, promoting calmness and reducing stress. On the other hand, activities that stimulate gamma waves, such as solving complex problems or learning new skills, can enhance cognitive performance and creativity.

Advances in Neuroscience: Mapping the Brain's "Electric Landscape"

Modern neuroscience has made remarkable strides in mapping the brain's electrical activity, often referred to as its "electric landscape." Techniques such as electroencephalography (EEG) and magnetoencephalography (MEG) allow scientists to record and analyze brainwaves in real time, providing a window into the brain's dynamic functioning. These tools have been invaluable for understanding the neural basis of consciousness, emotion, and behavior.

Advances in brain mapping have also revealed the importance of neural networks, clusters of interconnected neurons that work together to perform specific tasks. The default mode network, for example, is active during introspective activities like daydreaming and self-reflection. In contrast, the salience network helps the brain prioritize attention to important stimuli. By studying these networks and their electrical activity, scientists are uncovering the mechanisms that drive human thought and behavior.

In addition to research, brain-mapping technologies have practical applications in medicine and mental health. For instance, brain stimulation techniques like transcranial magnetic stimulation (TMS) and deep brain stimulation (DBS) use targeted electrical pulses to treat conditions like depression, Parkinson's disease, and epilepsy. These therapies work by modulating the brain's electrical activity, restoring balance to dysfunctional neural circuits.

Understanding the brain's electrical landscape also has implications for personal growth and self-optimization. Emerging tools like neurofeedback allow individuals to train their brainwaves by providing real-time feedback on their electrical activity. By learning to control their brainwave patterns, people can enhance focus, reduce anxiety, and improve emotional regulation.

The study of brainwaves and the electric mind not only deepens our understanding of human consciousness but also empowers individuals to harness their bioelectric potential. By exploring the interplay between electrical activity and mental states, we open new possibilities for improving well-being, expanding creativity, and unlocking the brain's full capacity. Through the intricate rhythms of bioelectricity, the brain reveals its power as both a biological organ and the seat of human experience.

Chapter 10: Electricity and Human Potential

Exploring Electricity's Role in Creativity, Problem-Solving, and Meditative States

The human brain's electrical activity is not only the foundation of thought and emotion but also a driving force behind creativity, problem-solving, and deep meditative states. The rhythmic oscillations of brainwaves, ranging from delta to gamma frequencies, create a dynamic environment where the mind can shift between intense focus, imaginative exploration, and profound relaxation.

Creativity often emerges during states where the brain's alpha and theta waves dominate. In these relaxed yet alert states, the brain is less constrained by rigid patterns of thought, allowing for the free association of ideas. For example, alpha waves often peak during moments of insight or when solving problems that require lateral thinking. Similarly, theta waves are associated with bursts of creative inspiration, often experienced during daydreaming or while engaging in repetitive, meditative activities like walking or drawing.

Problem-solving, on the other hand, benefits from the interplay of slower alpha waves and faster beta or gamma waves. When confronted with a complex challenge, the brain switches between relaxed states, where ideas are formed, and high-frequency activity, where these ideas are critically evaluated and refined. This dynamic interplay allows humans to approach problems with both creativity and precision, often leading to breakthroughs in understanding.

Meditative states, which are characterized by slower brainwave activity, also rely on the brain's electrical rhythms. Practices like mindfulness and focused breathing have been shown to increase alpha and theta activity, promoting relaxation and reducing stress. These meditative states are not just calming; they also enhance cognitive clarity, emotional resilience, and the ability to remain focused under pressure. By cultivating practices that optimize brainwave patterns, individuals can tap into their bioelectric potential to unlock new levels of creativity, problem-solving ability, and inner peace.

The Potential of Electrical Brain Stimulation to Enhance Cognition or Treat Disorders

Recent advancements in neuroscience have revealed that targeted electrical stimulation can enhance brain function and treat a variety of neurological and psychological disorders. Techniques such as transcranial magnetic stimulation (TMS), transcranial direct current stimulation (tDCS), and deep brain stimulation (DBS) have shown promise in improving cognition, alleviating mental health conditions, and restoring function in patients with brain injuries.

TMS, for example, uses magnetic fields to stimulate specific regions of the brain, enhancing neural activity in areas associated with attention, memory, and decision-making. This non-invasive technique has been used to treat depression and anxiety, with some studies suggesting that it can also improve cognitive performance in healthy individuals.

Similarly, tDCS delivers low electrical currents to the scalp, modulating brain activity in targeted regions. Research has shown that tDCS can enhance skills like problem-solving, language acquisition, and motor coordination. Athletes, students, and professionals are increasingly exploring tDCS as a tool for gaining a competitive edge, though its long-term effects are still being studied.

DBS, which involves implanting electrodes directly into the brain, has been life-changing for individuals with severe conditions like Parkinson's disease and obsessive-compulsive disorder. By delivering precise electrical pulses to specific brain regions, DBS can alleviate symptoms and improve quality of life. These interventions highlight the power of bioelectricity in restoring balance to neural circuits and enhancing brain function.

While electrical brain stimulation holds immense potential, it also raises ethical and practical questions. Should cognitive enhancement be available to everyone, or only for therapeutic use? How do we balance the benefits of stimulation with the risks of overuse or unintended effects? As these technologies evolve, they promise to expand the boundaries of human potential. Still, they also challenge us to consider how best to use them responsibly.

Speculations About How Bioelectricity Shapes Individuality and Personality

The unique patterns of electrical activity in the brain are not just reflections of thought and emotion; they may also play a role in shaping individuality and personality. Every person's brain generates a distinct bioelectric "signature," influenced by genetics, environment, and life experiences. These patterns govern not only how we perceive the world but also how we respond to it, forming the basis of traits like extroversion, creativity, and resilience.

For instance, studies have shown that differences in alpha wave activity can be linked to variations in creativity and problem-solving styles. Similarly, individuals with higher levels of gamma wave activity tend to exhibit heightened cognitive integration, allowing them to process information from multiple perspectives. These differences suggest that bioelectricity contributes to the diversity of human thought and behavior.

Speculative theories propose that bioelectric patterns may even underlie aspects of personal identity, influencing preferences, habits, and decision-making. The interplay of electrical activity across brain regions could create unique neural "maps" that guide how individuals interpret experiences and form relationships. Understanding these bioelectric underpinnings could lead to breakthroughs in personalized medicine, enabling interventions tailored to an individual's neural profile.

On a broader level, bioelectricity challenges traditional notions of identity by revealing how deeply connected we are to our environment. The electrical activity in our brains is influenced by external factors such as light, sound, and even the electromagnetic fields of the Earth. This interconnection suggests that individuality is not a fixed property but a dynamic process shaped by the constant exchange of energy and information between ourselves and the world around us.

By exploring how bioelectricity shapes personality and individuality, we gain a deeper understanding of what makes each person unique. This knowledge can inspire greater empathy, as we recognize the shared electrical foundation of human experience, while also empowering individuals to harness their bioelectric potential for personal growth and self-discovery. In the intricate dance of electrical signals within the brain, we find the essence of human potential, creativity, and connection.

Chapter 11: Electricity and the Life Force

Perspectives from Ancient Traditions: Electricity as Life Force

Across cultures and centuries, ancient traditions have recognized a mysterious, vital energy flowing through all living things. Known by different names—*chi* in Chinese philosophy, *Prana* in Indian yogic traditions, and *ki* in Japanese practices—this life force has often been associated with vitality, health, and the essence of being. While these traditions lacked the scientific understanding of bioelectricity, their descriptions of an invisible energy coursing through the body resonate with modern discoveries about the electrical systems that sustain life.

In Traditional Chinese Medicine (TCM), *chi* is said to flow through pathways called meridians, connecting different parts of the body. Techniques like acupuncture aim to balance this energy flow, promoting health and well-being. From a modern perspective, acupuncture has been shown to influence the nervous system, possibly affecting the body's bioelectric signals and promoting self-regulation. Similarly, in yoga, the flow of *Prana* is associated with breath control and meditation, practices that are now known to influence brainwave activity and autonomic nervous system balance.

These ancient concepts may have been early attempts to describe the bioelectric phenomena underlying life. Though the vocabulary differs, the parallels between the life force of ancient traditions and modern bioelectricity suggest a timeless awareness of the energy that animates living beings.

Bioelectromagnetism and the Concept of a "Biofield"

Modern science has uncovered that the human body not only generates electrical signals but also creates electromagnetic fields. This phenomenon, known as bioelectromagnetism, is evident in the measurable fields produced by the heart and brain. For instance, the heart's electrical activity creates a magnetic field detectable several feet from the body. In contrast, the brain's activity generates patterns that can be recorded using magnetoencephalography (MEG).

These discoveries have given rise to the concept of a "biofield," an energy field surrounding and permeating the body. While the biofield is not universally accepted in the scientific community, proponents argue that it represents the interplay of the body's electromagnetic activity with its surrounding environment. This concept has been explored in complementary and alternative medicine, where practices like Reiki and therapeutic touch are believed to influence the biofield to promote healing and balance.

From a speculative standpoint, the biofield could serve as a bridge between physical and energetic dimensions of life. Just as electrical signals within the body regulate its functions, the biofield may represent a larger, interconnected system of energy that extends beyond the boundaries of the physical form. This idea aligns with ancient traditions that view the human body as part of a larger energetic system interconnected with nature and the cosmos.

Speculations about Electricity as a Bridge between Science and Spirituality

Electricity, as a fundamental force of nature, occupies a unique position in the exploration of science and spirituality. On one hand, it is a measurable, physical phenomenon that powers life and technology. On the other, its invisible and pervasive nature invites metaphysical speculation about its role in connecting the physical and spiritual realms.

One speculative idea is that bioelectricity could serve as a medium for consciousness, bridging the gap between the material brain and the intangible experiences of thought, emotion, and awareness. This aligns with theories in neuroscience that suggest consciousness arises from the coordinated electrical activity of neural networks. Extending this idea, some propose that the biofield or electromagnetic activity of the brain might interact with larger fields, such as the Earth's electromagnetic field or the Schumann Resonance, creating a connection between individual consciousness and the collective or universal.

Ancient traditions often describe life force energy as a link between the body and the soul. In this light, bioelectricity could be interpreted as the scientific counterpart to these spiritual concepts, representing the energy that animates and connects all living things. For example, the heart's electromagnetic field, which extends beyond the body, could symbolize the interconnectedness of human beings, resonating with ideas of unity and shared consciousness found in many spiritual teachings.

Speculatively, electricity could also be seen as the foundation of spiritual transformation. Practices like meditation and prayer are known to alter brainwave patterns and electrical activity, producing states of calm, insight, and transcendence. These shifts in bioelectric rhythms might correspond to changes in perception and awareness, enabling individuals to access deeper aspects of their consciousness.

While these ideas remain speculative, they highlight the potential of electricity as a unifying force that bridges the physical and the spiritual. By exploring the connections between ancient wisdom, bioelectricity, and the mysteries of the human experience, we can deepen our understanding of what it means to be alive, conscious, and interconnected with the world around us.

In the end, electricity may not only illuminate the workings of the body and mind but also offer a glimpse into the profound and timeless mysteries of the soul.

Chapter 12: Ancient Cultures and the Electric Connection

Egyptians: Electric Fish and Healing Practices; the Concept of the "Ka" as Life Force

The ancient Egyptians were among the first to explore the mysterious properties of electricity, even if they lacked the scientific terminology to describe it. Their knowledge was rooted in observation and practical application. Electric fish, such as the electric catfish (*Malapterurus electricus*), were known to deliver powerful shocks when touched. Egyptians referred to these fish as the "Thunderers of the Nile" and believed their electric discharges held healing properties. Historical accounts suggest that they used these fish to treat pain, such as headaches and joint issues, by placing the fish on the affected area.

Beyond their practical use of electric fish, the Egyptians also had a philosophical framework for understanding life force energy. They believed in the "Ka," a spiritual essence or vital force that sustained life and connected the physical and spiritual realms. While the Ka was not explicitly linked to electricity, it shares conceptual similarities with modern ideas about bioelectric energy, serving as a bridge between the material body and the unseen forces that animate it.

Chinese: Qi and Acupuncture; Parallels to Bioelectric Pathways

In Chinese philosophy, *Qi* (or Chi) is described as the vital energy that flows through all living things, sustaining health and harmony. This energy is believed to travel along pathways called meridians, which connect different parts of the body. Practices such as acupuncture and Tai Chi are designed to balance and enhance the flow of Qi, promoting physical and emotional well-being.

Modern research into acupuncture has revealed parallels between meridians and the body's nervous system. Studies have shown that acupuncture points often correspond to areas with high concentrations of nerves or bioelectric activity. Stimulating these points may influence the body's electrical signals, potentially affecting pain perception, stress response, and immune function. While Qi remains a metaphysical concept, the mechanisms of acupuncture suggest that the ancient understanding of energy pathways aligns with the bioelectric networks we now recognize in the human body.

Greeks: Thales and Static Electricity; Early Theories of the Soul as an Energetic Force

The ancient Greeks were pioneers in observing and documenting electrical phenomena. Thales of Miletus, a philosopher from the 6th century BCE, is credited as one of the first to study static electricity. He discovered that rubbing amber with fur produced an attractive force, a phenomenon we now know as electrostatics. Although he could not explain this effect scientifically, Thales' curiosity laid the groundwork for later explorations of electricity.

Greek philosophers also speculated about the nature of the soul and its connection to energy. Plato and Aristotle described the soul as an animating force vital to the functioning of the body. While their ideas were more metaphysical than scientific, they resonate with modern notions of bioelectricity as the energy that sustains life and enables thought, emotion, and movement.

Indians: Prana and Chakras as Possible Analogs for Bioelectric Energy

In Indian traditions, *Prana* is considered the life force that flows through the body, similar to Qi in Chinese philosophy. Prana is said to travel through energy channels called nadis, which intersect at focal points known as chakras. Each chakra is associated with specific physical, emotional, and spiritual functions, and practices like yoga and breathwork are designed to enhance the flow of Prana.

The concept of chakras aligns intriguingly with the body's bioelectric systems. For example, the locations of the major chakras often correspond to clusters of nerve plexuses or endocrine glands. The root chakra, located at the base of the spine, aligns with the sacral plexus, while the crown chakra at the top of the head is near the brain's pineal gland. Modern science suggests that practices like deep breathing and meditation can influence brainwave activity and autonomic nervous system balance, offering a physiological basis for the effects described in ancient texts.

Native Americans and Mesoamericans: Reverence for Lightning as a Creative Force

Many Native American and Mesoamerican cultures revered lightning as a symbol of creation and divine power. For example, the Navajo considered lightning to be a manifestation of supernatural energy associated with the deity Thunderbird. Similarly, the Mayans believed that lightning was a tool of the gods, used to shape the Earth and bring life to its inhabitants.

These cultures recognized the transformative power of lightning, not just as a destructive force but also as a source of renewal and fertility. This reverence echoes modern theories about the role of lightning in the origin of life. Lightning's electrical energy, striking Earth's primordial atmosphere, may have catalyzed the formation of amino acids and other building blocks of life, making it both a literal and symbolic creator.

Electric Rays and Healing: Roman and Islamic use of Electric Fish for Medical Treatment

The practical use of electricity for healing extended into Roman and early Islamic medicine. Roman physicians such as Scribonius Largus described the therapeutic use of electric rays, or torpedo fish, to treat pain and ailments like gout and headaches. Patients would place their hands or feet on the fish, allowing its electric shocks to numb the affected area and provide relief.

Islamic scholars preserved and expanded upon these practices, incorporating them into their medical texts. The use of electric fish represents one of the earliest examples of electrotherapy, demonstrating an intuitive understanding of electricity's potential to influence the human body. These methods predate the modern use of electrical stimulation in pain management and physical therapy, highlighting the timelessness of electricity as a tool for healing.

How these Ancient Beliefs and Practices Align with Modern Discoveries

The parallels between ancient traditions and modern science reveal a profound continuity in humanity's exploration of electricity and life. From the Egyptians' concept of the Ka to the Chinese understanding of Qi and the Indians' description of Prana, these ancient ideas capture the essence of bioelectricity as a life-sustaining force.

Modern research has validated many of the mechanisms underlying these practices, such as the influence of acupuncture on nerve function or the alignment of chakras with bioelectric and physiological centers in the body. Even speculative ideas about energy fields and life force resonate with the emerging science of biofields and the measurable electromagnetic activity generated by living organisms.

By bridging ancient wisdom with contemporary science, we gain a richer understanding of electricity's role in life. These cultural perspectives remind us that the search for meaning and connection is universal and timeless, grounded in the recognition of an invisible force that animates and unites all living things. Whether viewed through the lens of spirituality or science, electricity continues to inspire curiosity, innovation, and reverence for the mysteries of existence.

Part 5: The Future of Electricity and Life

Chapter 13: Bioelectricity in Medicine and Technology

Advances in Bioelectric Engineering: Artificial Organs and Regenerative Medicine

The convergence of bioelectricity and modern engineering is revolutionizing medicine, particularly in the development of artificial organs and regenerative therapies. As scientists deepen their understanding of how bioelectric signals guide cellular processes, they are designing technologies that replicate or enhance these natural signals to restore function and promote healing in the human body.

One of the most groundbreaking advancements is the creation of bioengineered tissues and artificial organs that use electrical cues to function. Bioelectric scaffolds, for instance, are synthetic or biological frameworks seeded with stem cells and programmed to generate subtle electric fields. These fields mimic the body's natural signals, guiding the cells to grow, differentiate, and organize into functioning tissues. This approach has shown promising results in regenerating damaged skin, muscles, and even cardiac tissue. For example, researchers have successfully used bioelectric scaffolds to repair heart damage by stimulating the growth of new, functional heart muscle cells.

The development of artificial organs, such as bioelectric kidneys, lungs, or livers, represents another frontier. These technologies aim to restore full organ function by incorporating sensors and stimulators that imitate the bioelectric communication found in natural organs. The ultimate goal is to create fully functional, transplantable organs grown from a patient's own cells. This could reduce the risk of rejection and eliminating the need for donor waiting lists.

In regenerative medicine, electrical stimulation is already being used to accelerate tissue repair. Studies have shown that applying specific electric fields to damaged tissues can encourage cell migration, division, and growth. This technique has been particularly successful in treating chronic wounds like diabetic ulcers. It is now being adapted for nerve and bone regeneration. The integration of bioelectric engineering with stem cell therapies offers new hope for conditions that were once considered untreatable.

Using Electricity to Combat Disease

The therapeutic potential of electricity extends beyond regeneration to the treatment of diseases, offering innovative solutions for conditions ranging from cancer to neurological disorders. One of the most promising applications is in cancer therapy, where bioelectric treatments are being explored to target and destroy tumors.

Electric field therapies, such as Tumor Treating Fields (TTFields), use low-intensity alternating electric fields to disrupt the division of cancer cells. By interfering with the cell's mitotic process, these fields prevent tumors from growing and spreading while leaving healthy cells largely unaffected. Clinical trials have demonstrated the effectiveness of TTFields in treating aggressive cancers like glioblastoma, providing patients with a non-invasive and targeted therapy that complements conventional treatments like chemotherapy and radiation.

In nerve repair, electricity has proven to be a powerful tool for regenerating damaged neural pathways. Techniques like electrical nerve stimulation help activate growth factors that encourage nerve fibers to regenerate and reestablish connections. This approach has been used successfully in patients recovering from spinal cord injuries or peripheral nerve damage, restoring movement, sensation, and functionality.

Electrical stimulation also holds promise in treating neurological disorders like Parkinson's disease, epilepsy, and depression. Deep brain stimulation (DBS), a therapy that delivers targeted electrical pulses to specific regions of the brain, has already transformed the lives of thousands of patients with Parkinson's. Similarly, transcranial magnetic stimulation (TMS) and transcranial direct current stimulation (tDCS) are being used to alleviate symptoms of depression and improve cognitive function by modulating neural activity.

Beyond these applications, bioelectric therapies are being explored for conditions such as chronic pain, heart arrhythmias, and immune system dysfunction. By harnessing electricity's ability to influence cellular behavior and communication, scientists are developing therapies that are more precise, effective, and less invasive than traditional medical interventions.

Emerging Technologies Inspired by Biological Electricity

The natural bioelectric processes observed in living organisms are inspiring a new wave of technologies that mimic or interact with biological systems. These innovations, often referred to as bio-inspired or biomimetic technologies, promise to revolutionize fields ranging from medicine to sustainable energy.

One of the most exciting areas of research is bioelectric implants that interface seamlessly with the body's nervous system. Brain-computer interfaces (BCIs), for example, use electrical signals from the brain to control man made devices, such as robotic limbs or communication tools for people with paralysis. Companies like Neuralink are pushing the boundaries of this technology, aiming to create direct brain-machine connections that could restore mobility, speech, and independence for individuals with neurological injuries or diseases.

Bioelectric sensors are another emerging technology inspired by the body's natural ability to detect and respond to electrical signals. These sensors can monitor biological functions in real-time, such as heart activity, brainwaves, or glucose levels, providing personalized data to guide treatment. Wearable devices equipped with bioelectric sensors are already transforming healthcare, allowing individuals to track their health and respond proactively to changes in their body.

In the energy sector, researchers are exploring bioelectric solutions inspired by electric-producing organisms like bacteria and electric fish. Microbial fuel cells, for example, harness the bioelectric activity of bacteria to generate clean energy. By breaking down organic matter, these microbes release electrons that can be captured and converted into electricity. This technology has the potential to create sustainable energy solutions for communities and industries.

The concept of bioelectricity is also driving innovations in environmental monitoring. Bio-sensing plants equipped with electrical detection systems can be used to monitor soil quality, water pollution, or climate changes. These plants essentially act as living sensors, providing real-time data about the health of ecosystems while reducing the need for expensive and invasive testing methods. The merging of bioelectricity with technology represents a profound shift in how we approach human health, environmental sustainability, and technological advancement. By looking at the natural electrical systems that have evolved over billions of years, scientists and engineers are uncovering new ways to solve complex challenges.

This rapidly evolving field reminds us that the power of electricity extends far beyond wires and machines—it is woven into the fabric of life itself. By understanding and harnessing this energy, we are not only advancing science and technology but also unlocking new possibilities for healing, growth, and connection. The future of electricity and life holds the promise of a world where technology and biology work in harmony, driven by the same invisible force that has shaped life from its very beginning.

Chapter 14: Electricity and Artificial Intelligence

How Studying Human Bioelectricity Informs AI and Robotics

The study of human bioelectricity has provided invaluable insights into the development of artificial intelligence (AI) and robotics, helping scientists and engineers design systems that emulate human thought, movement, and adaptability. At the core of this inspiration is the brain, a biological marvel that processes electrical signals to create thought, decision-making, and action with remarkable efficiency. Understanding how the brain's bioelectric systems work has become a blueprint for creating intelligent machines.

Artificial intelligence draws heavily from the concept of neural networks, which are modeled after the brain's own network of neurons. In the human brain, each neuron communicates using electrical signals transmitted through action potentials. This system of interconnected, firing neurons creates dynamic patterns of activity that allow for learning, memory, and problem-solving. In AI, artificial neural networks replicate this concept through layers of digital "neurons" that process information, learn from data, and improve over time through feedback loops. While digital models lack the complexity and bioelectric basis of human cognition, the parallels are unmistakable. Each system relies on electrical signals to generate intelligent behavior.

Bioelectricity also influences the design of robotics, particularly in the development of responsive, adaptive systems. Just as bioelectric signals control muscle contractions in humans, electrical impulses in robotics control motors and actuators to enable movement. Researchers are increasingly looking at how bioelectric systems can inspire robots that react to their environment with lifelike precision. For instance, engineers are developing soft robotics that use electrical signals to mimic the flexibility and dexterity of biological muscles, paving the way for robots capable of fine motor tasks, such as surgery or assisting individuals with disabilities.

The emerging field of neuromorphic computing takes this inspiration even further. Neuromorphic systems aim to recreate the bioelectric activity of the human brain using specialized hardware that mimics the way neurons process and transmit electrical signals. These systems are not only faster and more energy-efficient than traditional computing but also capable of handling ambiguous and unstructured data, much like the human brain. This innovation represents a step toward machines that think and learn in ways that are both biologically inspired and deeply efficient.

Speculative Futures: Integrating Human Bioelectricity with Machines

The potential to integrate human bioelectricity with machines represents one of the most profound and speculative frontiers in science and technology. As research into bioelectricity and AI advances, scientists are exploring ways to merge the electrical systems of living organisms with artificial systems, creating interfaces that blur the line between biology and technology.

Brain-computer interfaces (BCIs) are already making this vision a reality. By detecting and interpreting the brain's electrical signals, BCIs enable direct communication between the human brain and machines. For example, individuals with paralysis can use BCIs to control robotic limbs or computers simply by thinking about specific actions. This technology has the potential to restore mobility, independence, and communication for people with neurological injuries or diseases.

Looking further into the future, the integration of human bioelectricity with machines could enhance not only function but also human potential. Imagine a world where neural implants allow humans to access vast stores of information instantaneously, communicate telepathically through electrical signals, or even augment their cognitive abilities. By interfacing directly with the brain's electrical networks, machines could help humans overcome biological limitations, opening doors to levels of intelligence, memory, and creativity that were once unimaginable.

This merging of biology and technology also raises philosophical and ethical questions. If machines can interface seamlessly with the human brain, where does the boundary between human and machine lie? Would enhancing cognition or physical abilities through bioelectric interfaces change the essence of what it means to be human? These questions challenge our perceptions of identity, consciousness, and the role of technology in our lives.

Speculatively, integrating bioelectricity with machines could extend beyond the brain. The body's natural electrical systems might one day be used to power wearable devices, sensors, or even implants that monitor health in real time. Bioelectric signals could control smart prosthetics that feel like natural limbs, respond to subtle nerve impulses, and adapt dynamically to an individual's needs. In such a future, machines would no longer be external tools but integrated extensions of the human body, enhancing our capabilities while maintaining harmony with our biology.

At the farthest edges of speculation, some propose that bioelectricity could serve as a bridge to collective intelligence. Suppose human brains, with their electrical activity, could be networked together via advanced interfaces. In that case, it might be possible to share experiences, thoughts, and problem-solving processes on an unprecedented scale. This vision aligns with concepts of "hive minds" or collective consciousness, often explored in science fiction but grounded in the real understanding of electricity's role in communication and connection.

While much of this remains speculative, the trajectory is clear: the integration of bioelectricity and artificial intelligence has the potential to transform human life in ways both practical and profound. By understanding and harnessing the principles of bioelectricity, we are not just building smarter machines—we are redefining the possibilities of the human experience, merging biology and technology into a new era of human evolution.

Conclusion

Revisiting the Journey

The story of electricity is, at its heart, the story of life itself. Our journey began billions of years ago in the volatile and storm-laden atmosphere of early Earth, where bolts of lightning likely provided the energy to spark the first building blocks of life. From that humble beginning, electricity has remained an invisible yet fundamental force, guiding the evolution of organisms, shaping the nervous systems of animals, and powering the complexity of the human brain.

Through this exploration, we have seen how bioelectricity serves as the foundation for communication, growth, and healing across all living systems. Plants rely on electrical signals to sense and respond to their environment. At the same time, animals evolved intricate nervous systems to move, adapt, and survive. In humans, bioelectricity powers our thoughts, emotions, and even our capacity for self-awareness. From the generation of brainwaves to the rhythmic beating of the heart, electricity is woven into the very fabric of our existence.

At the same time, this journey has revealed how deeply interconnected humans are with the electrical forces of the Earth and the universe. Ancient cultures, without the tools of modern science, intuited the existence of a life force, a current of energy that sustains all living beings. Whether through the Egyptian Ka, Chinese Qi, or Indian Prana, these traditions echo the modern understanding of bioelectricity and its role in vitality and consciousness.

The mysteries of the mind, too, highlight the profound possibilities of electricity. The brain, a biological conductor of electrical energy, enables human creativity, reasoning, and emotional experience. As we've explored, brainwave frequencies can shift states of consciousness, bridging the gap between science and spirituality and offering insights into the essence of being human.

How Understanding Electricity can Transform Medicine, Technology, and Philosophy

As we stand at the intersection of science, technology, and philosophy, understanding electricity opens doors to a future where human potential can be transformed. Advances in bioelectric medicine are already revolutionizing healthcare, offering solutions for healing wounds, repairing nerves, and even regenerating tissues and organs. The ability to influence and guide the body's electrical signals brings new hope for treating conditions once deemed untreatable, from paralysis to chronic pain.

In technology, the lessons of bioelectricity are inspiring innovations that mimic the elegance of living systems. Brain-computer interfaces are bridging the gap between humans and machines, creating pathways for restoring mobility, enhancing cognition, and even redefining the limits of communication. Emerging technologies, like bioelectric sensors and energy systems inspired by nature, promise solutions for sustainability and human health, reflecting a deeper harmony between biology and engineering.

On a philosophical level, the study of bioelectricity invites us to reconsider our understanding of life and consciousness. If electrical signals underpin thought, memory, and identity, what does that mean for the essence of self? How do we reconcile our physical bodies with the immaterial experiences of awareness and connection? These questions challenge us to see electricity not just as a force of nature but as a bridge between the tangible and the profound.

Final Thoughts

At the conclusion of this journey, one question remains: Is electricity the essence of life? The evidence suggests that it is not merely a tool but a unifying force that connects all living things, from the simplest organisms to the most complex beings. Electricity powers the processes that sustain us, yet it also gives rise to experiences of creativity, thought, and connection that define what it means to be alive.

Perhaps electricity is more than just a physical phenomenon—it may be the thread that ties together the biological, the energetic, and the spiritual. It animates the cells in our bodies, creates the signals that allow us to think and feel, and links us to the world through unseen yet measurable currents. Whether through the Schumann Resonance aligning with brainwaves, the electrical mysteries of the mind, or the life force described by ancient cultures, electricity seems to connect the material and the immaterial, the seen and the unseen.

The story of electricity challenges us to see the world differently. It invites us to appreciate the beauty of life's invisible currents, the harmony of our bioelectric systems, and the vast possibilities for transformation when we learn to understand and harness this fundamental force. Suppose electricity is indeed the essence of life. In that case, it reminds us that we are part of something far greater than ourselves—an interconnected flow of energy that unites all living things.

By exploring electricity, we do more than uncover the workings of biology and technology; we touch the very heart of existence. It is a force that sustains, heals, inspires, and connects—an invisible current that pulses through us all. In understanding it, we may not only unlock new possibilities for the future but also discover what it truly means to be alive.

The emerging field of neuromorphic computing takes this inspiration even further. Neuromorphic systems aim to recreate the bioelectric activity of the human brain using specialized hardware that mimics the way neurons process and transmit electrical signals. These systems are not only faster and more energy-efficient than traditional computing but also capable of handling ambiguous and unstructured data, much like the human brain. This innovation represents a step toward machines that think and learn in ways that are both biologically inspired and deeply efficient.

www.ingramcontent.com/pod-product-compliance
Lightning Source LLC
Chambersburg PA
CBHW070420230526
45471CB00006B/2901